Mid- and Long-Term Prospects for Human Spaceflight

Mitigating the Gaps

A Report of the CSIS Space Initiatives

AUTHORS
Vincent Sabathier
G. Ryan Faith
Ashley Bander

February 2009

CSIS | CENTER FOR STRATEGIC & INTERNATIONAL STUDIES

About CSIS

In an era of ever-changing global opportunities and challenges, the Center for Strategic and International Studies (CSIS) provides strategic insights and practical policy solutions to decisionmakers. CSIS conducts research and analysis and develops policy initiatives that look into the future and anticipate change.

Founded by David M. Abshire and Admiral Arleigh Burke at the height of the Cold War, CSIS was dedicated to the simple but urgent goal of finding ways for America to survive as a nation and prosper as a people. Since 1962, CSIS has grown to become one of the world's preeminent public policy institutions.

Today, CSIS is a bipartisan, nonprofit organization headquartered in Washington, DC. More than 220 full-time staff and a large network of affiliated scholars focus their expertise on defense and security; on the world's regions and the unique challenges inherent to them; and on the issues that know no boundary in an increasingly connected world.

Former U.S. senator Sam Nunn became chairman of the CSIS Board of Trustees in 1999, and John J. Hamre has led CSIS as its president and chief executive officer since 2000.

CSIS does not take specific policy positions; accordingly, all views expressed herein should be understood to be solely those of the author(s).

Cover photo credits: NASA, top left; JAXA/NHK, top right; U.S. Air Force, bottom left; Arianespace, bottom right.

Library of Congress Cataloguing-in-Publication Data
CIP Information available on request.
ISBN 978-0-89206-571-4

Center for Strategic and International Studies
1800 K Street, NW, Washington, DC 20006
Tel: (202) 775-3119
Fax: (202) 775-3199
Web: www.csis.org

CONTENTS

MID- AND LONG-TERM PROSPECTS FOR HUMAN SPACEFLIGHT

MITIGATING THE GAPS

Vincent Sabathier, G. Ryan Faith, and Ashley Bander

Introduction

The CSIS Space Initiatives project promotes global space exploration as we move into the twenty-first century. The next decade, ushered in by the new U.S. president and his new administration, will be crucial to space exploration. Indeed, during the next decade, the Space Shuttle will have to be retired, the International Space Station (ISS) made a visible success, and humans returned to the moon. In fact, the Government Accountability Office (GAO) recently identified retirement of the Space Shuttle as one of the 13 most urgent issues that the next administration will need to address.[1] CSIS's Space Initiatives believes that there will be no global space exploration without the right U.S. leadership. On the other hand, the United States intends to reestablish a human presence on the moon by 2019, the 50th anniversary of the Apollo 11 moon landing, this time creating a permanent, international presence. CSIS's Space Initiatives put together a select working group to provide objective, bipartisan, and pragmatic insights into this complex set of issues. The first working group meeting took place on November 14, 2008, and brought together representatives from the U.S. government, industry, international partners, and policy associations to define the foreseeable gaps and explore the consequences of these gaps for the future of human spaceflight. A second working group meeting, on December 8, 2008, evaluated scenarios for mitigating the gaps, keeping in mind budgetary, political, foreign policy, and technical objectives and constraints.[2] The current report represents the views of its authors and not the individual stances of working group members or of CSIS.

Background

"The Gap" in U.S. human spaceflight most literally refers to the interval between retirement of the Space Shuttle, currently scheduled for September 2010, and the initial operational capability (IOC) of the first elements of the Constellation Program, currently expected in 2015. The

[1] Government Accountability Office (GAO), "2009 Congressional and Presidential Transition: Urgent Issues Overview," http://www.gao.gov/transition_2009/urgent/.

[2] See the appendix for meeting agendas. The presentations are available on the Space Initiatives Web site, http://www.csis.org/space/.

proposal for retiring the shuttle and moving to a new system, outlined in the 2005 Exploration Systems Architecture Study (ESAS),[3] addressed current and future space transportation needs of the nation extraordinarily well. Sound both technically and financially, it limited the gap, with the Constellation system ready in 2013, or as early as 2011 with extra funding. Ares I and V would reuse existing shuttle elements, and the Orion crew capsule was based on proven technologies from past programs. Further, the so-called 1½ architecture would provide early capability to get both humans and cargo to—and from—the International Space Station. The combination of a human-rated Ares I and a super-heavy Ares V would be ideal for the moon. In addition, Ares V would, by design, provide flexibility and interesting capability beyond the moon and beyond human spaceflight.

This approach enjoyed support from industry and constituents and was the ideal solution when it was proposed, which makes it a difficult plan to alter. Changes since the plan was created, however, make it necessary to do so. For one, the National Aeronautics and Space Administration (NASA) has consistently received inadequate funding for Constellation development. Second, that development has moved from being shuttle derived to something more like "shuttle flavored." The Orion crew capsule remains a robust design, providing safe and theoretically interoperable crew transportation—as well as up- and down-mass for cargo—with a comparatively short time for development. However, the Ares I launcher, which is at the center of the Constellation Program, has experienced a number of changes, including the switch to a five-segment booster and the J2X engine—new systems to develop, with the attendant added costs, delays, and uncertainties. Additionally, the delayed IOC has increased workforce transition problems. As things currently stand, Ares I/Orion is expected to begin operating, at the earliest, in 2015. Some studies, including a recent one from the Congressional Budget Office (CBO),[4] estimate that the system could be delayed as much as 18 months beyond the currently planned IOC date of March 2015, due in large part to cost growth beyond current Constellation reserves.

Extending the shuttle for additional missions could mitigate the gap only if additional funding is provided by Congress for those missions. Indeed, in the NASA Authorization Act of 2008,[5] the U.S. Congress directed that agency to take no steps that would preclude extending the shuttle past 2010. That same bill proposed an additional $1 billion to accelerate the development of Ares I/Orion. According to NASA, that $1 billion would shave nine months off the development time. Ares I/Orion cannot be accelerated much more than that, given both the technical limitations that surround the development of any new space system—particularly propulsion systems—and the

---

[3] National Aeronautics and Space Administration (NASA), *NASA's Exploration Systems Architecture Study: Final Report* (Washington, D.C.: NASA, November 2005), http://www.nasa.gov/mission_pages/constellation/news/ESAS_report.html.

[4] Kevin Eveker, "An Analysis of NASA's Plans for Continuing Human Spaceflight After Retiring the Space Shuttle," Congressional Budget Office (CBO), Washington, D.C., November 2008, http://www.cbo.gov/ftpdocs/98xx/doc9886/11-03-NASA_Letter.pdf.

[5] *National Aeronautics and Space Administration Act of 2008*, HR 6063, 110th Cong. http://thomas.loc.gov/cgi-bin/query/z?c110:H.R.6063:.

delays the program has already suffered due to budget constraints since it was initiated. As mentioned, many analyses of the Constellation Program—particularly Ares I—expect it to be delayed further.

The Three Gaps and the Need for Mitigation

This creates a significant gap in U.S. human spaceflight capability, five years or more during which the United States—and other ISS partners—will be wholly reliant on the Russian Soyuz to access the ISS.[6] However, there are two further gaps, one in ISS utilization and a second in plans for lunar exploration. The technical limitations of the Soyuz—for instance, lack of significant down-mass capability—threaten international partners' abilities to utilize the station for research. Without that research capability, many in ISS partner countries would see the station as a failure, including those politicians and policymakers that have a say in funding future manned space exploration. The U.S. spaceflight gap and the ISS utilization gap combine to create a complex problem: a moon gap. U.S. plans call for a human return to the moon by 2019 and the building of a permanent lunar base. As NASA administrator Michael Griffin noted in his speech at CSIS on November 1, 2005,[7] U.S. funding constraints will require the lunar base to be an international project. Current NASA budget plans indicate that the timely development of the Ares V vehicle and Altair lunar lander will have to be paid for by withdrawing U.S. support from the ISS in 2016. Although many in NASA and Congress have stated that the United States will remain involved in the ISS at least through 2020, widespread use of the 2016 date poses an additional threat to ISS development and utilization. European and Japanese partners are already wary of investing additional resources to develop items for a station that will not be supported in a few years. If they cannot show a success with the ISS, they will find it exceedingly difficult to obtain funding for further human spaceflight efforts, making their participation in a human lunar program quite unlikely.

Given that the United States is already relying on international cooperation on the lunar surface, an inability to attract partners in lunar exploration would severely limit the likelihood of achieving current U.S. plans, codified in both law and public statement, of resuming human space exploration beyond Low Earth Orbit (LEO). These three gaps threaten not just the next few years of the U.S. space program, but utilization of the ISS, international cooperation to return to the moon, and beyond—in short, the future of human spaceflight—and without mitigation, the situation will continue to deteriorate. Further delays in the development of Constellation will only increase the severity of these three gaps. Although some attention is rightly focused on securing or accelerating Constellation, options for mitigating the gap, for finding interim and innovative ways

---

[6] For more on access to the ISS, see Vincent Sabathier, G. Ryan Faith, and Alexandros Petersen, "Ensuring U.S. Access to the International Space Station," *CSIS Commentary*, August 15, 2008, http://www.csis.org/media/csis/pubs/080815_faith_georgia.pdf.

[7] See program for "International Cooperation on Space Exploration," http://www.csis.org/media/csis/events/051101_hsei.pdf.

to maintain an international human presence in space, must be developed. The risks to the future of human spaceflight posed by these three gaps and the associated foreign policy implications are simply too great.

Goals and Constraints of Mitigation

These gaps are fundamentally space transportation issues, with both national and international components, that will require international coordination to address. In mitigating the gaps, a number of goals and constraints have been identified. Three of the goals address the issue of the gaps directly—retire the shuttle in deference to a new system, make a success of the ISS, and return to the moon by 2019, in collaboration with international partners. Beyond those goals, mitigating the gap should help to engage new emerging international players to broaden human space exploration. Ultimately, mitigating the gap should renew and ensure U.S. leadership in space and beyond. Any mitigation, however, will take place in a constrained budgetary environment and must take into account workforce issues in the United States. Ares I is in the critical path for all of these goals—its delays become everyone's delays. In addition, current launch infrastructure in the United States and around the world will require support to remain operational. Given that the gap is fast approaching, any path for mitigating it must be pragmatic. Now may not be the time to propose funding for and reliance on only new launch systems and subsystems. Existing launchers and subsystems may open up potential mitigation options.

Options for Mitigation

Among the additional mitigation options discussed in the working group were Commercial Orbital Transportation Services, Capability D (COTS-D), either as planned or combining an expendable launch vehicle (ELV) and an Orion-lite, enhanced international cooperation, and the shuttle-derived DIRECT scenario.[8] The key to all of these scenarios is interoperability.[9] They all boil down to a crew capsule on a launch vehicle—the Dragon module on a Falcon 9, or some version of Orion atop an existing U.S. or international launcher or the shuttle-derived launcher envisioned in the DIRECT scenario. This degree of interoperability would remove the launch vehicle as the single point of failure, and open up new avenues for international cooperation in space exploration in LEO and beyond, both with existing partners and emerging spacefaring nations. Four main criteria are important in analyzing these options, in order of importance: safety, availability, affordability, and capability. Although crew safety is by far the most important factor, a launch abort system—a key concept in both the ESAS and the current Constellation plan—normalizes safety across the different scenarios in our analysis. Capability must be

---

[8] For more on the DIRECT scenario, see the presentation made to the working group at http://www.csis.org/space/.

[9] For more on interoperability, see G. Ryan Faith, Vincent Sabathier, and Lyn Wigbels, *Interoperability and Space Exploration* (Washington, D.C.: CSIS, June 2007), http://www.csis.org/component/option,com_csis_pubs/task,view/id,3901/type,1/.

considered for both immediate and long-term exploration; that is, LEO and beyond. The chart below compares the above options across factors of affordability, availability, and capability.

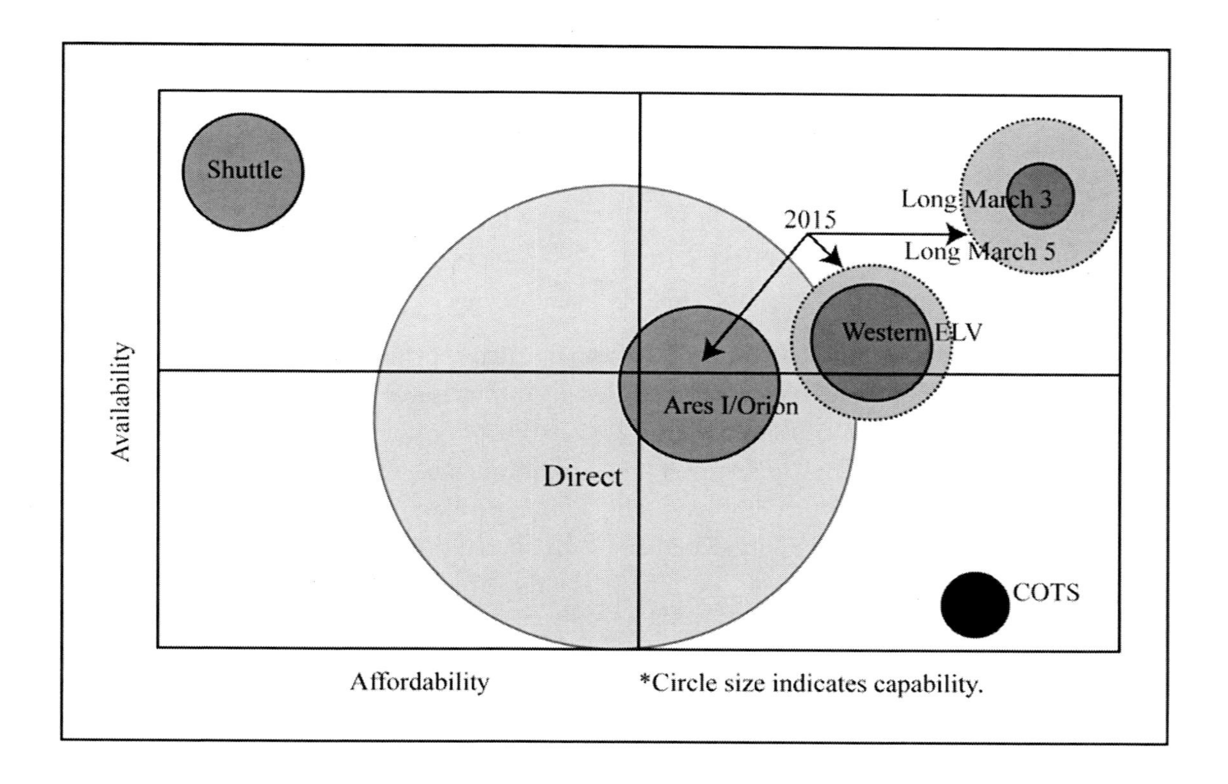

Three Budget Scenarios

Having identified several options for mitigating the gap, the working group analyzed some possible scenarios, given three budget environments: the status quo, which forecasts an essentially flat-line budget over the next 10 years; an ideal scenario in which NASA receives an additional $2 billion a year specifically for human spaceflight for the next 10 years; and a middle-ground scenario that assumes an increase of $1 billion a year for human space flight over the next 10 years. The scenarios analyzed had to consider how many more shuttle flights there will be, how many more years the ISS will be utilized, and how long elements of Constellation could be delayed.

In a status quo budgetary environment, the main question is "when will the various systems of the Constellation Program be ready?" which is, in part, related to "how much will the Constellation Program cost?" The current NASA timeline predicts that Ares I and Orion will begin operations in March 2015, with Ares V available in 2018. NASA's exploration budget includes $7 billion in Constellation reserves, to control the impact of cost growth. However, a recent CBO report predicted that, based on cost growth in past NASA projects, the development of Ares I/Orion is likely to be as much as $14 billion over budget, an increase that, without additional money for the program, would lead to a delay of as much as 18 months. That would mean that Ares I and Orion would not be ready until late in 2016. The delay in Ares I/Orion, as well as the cost growth that absorbed the existing Constellation budget reserves, would push Ares V well past 2018, which would, in turn, push a U.S. return to the moon well past 2019, leaving the United States to celebrate the Apollo anniversary with explanations of why, after half a century, we are still unable to get back to the moon. Additional cost increases beyond CBO projections, or unforeseen technical problems, of course, have the potential to delay Constellation even further. A status quo budget will not allow great flexibility for mitigating the gap. Proponents of the DIRECT scenario claim that the system envisioned there could be developed within the current budget. Although we do not have the resources to evaluate the DIRECT scenario, the proposal should be subjected to an independent technical, programmatic, and budgetary review, as it was when first envisioned.

With a budget increase of $2 billion a year for 10 years allocated specifically for human spaceflight, NASA could secure or even accelerate the current timetable for Constellation, as well as put money toward other options for mitigating the gap. Based on the CBO analysis mentioned above, $14 billion could be put toward securing the current timetable for Constellation. With the remaining money, additional shuttle flights could be put on the manifest to address the gap in the short term. Current hardware availability would allow for an additional four shuttle flights, which could spread out over two years for about $4 billion. The remaining $2 billion could then be used to invest in a COTS-D program that not only allows for the development of new systems, but also takes full advantage of existing launchers, both U.S. and international. This sort of budget increase would be an ideal scenario, but it would probably involve a larger increase for NASA as a whole since any increases in budget must also address programs such as earth science and aeronautics.

If NASA received a more realistic $1 billion a year for 10 years for human spaceflight programs, the focus on Constellation would shift to securing the ultimate lunar goals, while still providing funding that could go to additional shuttle flights or an expanded COTS-D effort. In this scenario, the Ares I and Orion budgets would be unchanged, risking potential delays. The development of Ares V by 2018, however, would be protected, with an additional $8 billion added to the program to make up for expected cost increases on Ares I/Orion. The remaining $2 billion could either be wholly invested in a revised and expanded COTS-D program, as outlined above, or split between the expanded COTS-D program and additional shuttle flights.

Recommendations

The options discussed above for mitigating the gaps are aimed, ultimately, at mitigating programmatic risk. Reaching interoperability among different launchers decouples the development risk associated with the crew capsule from the vulnerabilities of the launcher. Such interoperability also opens new avenues for international cooperation and leads to greater flexibility in human space exploration. These potential solutions would expand the options for putting people in space, expand all partners' abilities to utilize the ISS, and create new partnerships for the exploration of the moon and beyond. Just as the gap impacts more than just

U.S. human spaceflight efforts, the mitigation solutions have the potential to not only sustain, but also expand humankind's exploration of space.

Recommendation 1: Keep ambitious goals and fund them accordingly. Commit to utilizing the ISS through at least 2020 and to returning to the moon by 2019. A minimum $1 billion per year increase would secure Constellation's development with an eye toward lunar return, while developing a robust COTS-D program. Without this increase, U.S. human space exploration goals cannot be met. This increase must be a part of an overall budget increase for NASA that, among other things, maintains and improves earth observations capabilities.[10]

Recommendation 2: Engage international partners. Take advantage of the existing ISS intergovernmental agreement to reengage partners at the political level. Commit to the ISS at least through 2020; acknowledge and engage emerging spacefaring nations; and pursue interoperability in new spaceflight capabilities.

Recommendation 3: Validate the Constellation Program. Proceed with a quick and thorough independent review of the Ares I, with an eye toward the costs and opportunities for accelerating development. At the same time, secure or even speed up the development of Ares V and Orion. As part of this validation, alternate scenarios should be reassessed.

Recommendation 4: Revise and expand COTS-D. Additional funding should be provided for an expanded COTS-D program, at $2 billion, that would be open to existing launchers and international partnerships. This strengthened program would bring additional capacity online as soon as possible, creating a robust and interoperable fleet.

---

[10] For more on the importance of earth observations, see Lyn Wigbels, G. Ryan Faith, Vincent Sabathier *Earth Observations and Global Change: Why? Where Are We? What Next?* (Washington, D.C.: CSIS, July 2008), http://www.csis.org/media/csis/pubs/080725_wigbels_earthobservation_web.pdf.

Appendix: Working Group Agendas

November 14, 2008

2:00 pm –	Welcome Vincent Sabathier, CSIS
2:15 pm –	The Current Plans John Olson and Rich Leshner, NASA[11] Discussion
2:45 pm –	An Analysis of NASA's Plans for Continuing Human Spaceflight after Retiring the Space Shuttle Kevin Eveker, CBO Discussion
3:15 pm –	Break
3:30 pm –	The ISS Situation—A European perspective Emmanuel de Lipkowski, CNES Jürgen Drescher, DLR Andreas Diekmann, ESA The ISS Situation—A Japanese perspective Yoshinori Yoshimura, JAXA Discussion
4:00 pm –	The International Environment Ashley Bander, CSIS
4:15 pm –	Consequences of the Gaps ▪ Loss of U.S. Leadership by 2014 ▪ Dependence on Russia ▪ Underutilization of the ISS ▪ Delay of the Lunar Program
5:00 pm –	Close

[11] NASA participated to present information about NASA's current programs during the first meeting and associated question-and-answer session that followed. NASA did not participate in the dialogue to consider alternatives nor in the writing of the report during the second session. As such, this does not denote a NASA endorsement of the CSIS report, but rather captures NASA's effort to provide current factual information to the group.

December 8, 2008

2:00 pm –	Summary from the First Working Group: Characterization for the Three Gaps and the Need to Mitigate
2:15 pm –	Objectives and Constraints for Mitigating the Gaps
2:30 pm –	Gap Issues in the 2008 NASA Authorization Act Jeff Bingham, Senate Committee on Commerce, Science, and Transportation
2:45 pm –	Options Currently under Discussion

- Postpone Shuttle Retirement
- Accelerate Ares I/Orion
- COTS-D
  - —As currently planned
  - —ELV/Orion-lite
- Enhanced International Cooperation: Mitigating the Gaps while Building the Exploration Fleet
- DIRECT Scenario
  Steve Metschan, TeamVision Corporation

3:45 pm –	Break
4:00 pm –	Scenario Analysis
5:00 pm –	Conclusion

About the Authors

Vincent Sabathier is senior fellow and director for space initiatives at CSIS. He is also senior adviser to the SAFRAN group and consults internationally on aerospace and telecommunications. He is president of Als@tis, a telecommunications company he founded in Toulouse, France, in December 2004. Until September 2004, he served as the representative of the French Space Agency (CNES) in North America and as the attaché for space and aeronautics at the embassy of France in Washington, D.C. There, he focused on strengthening bilateral dialogue and cooperation with all branches of the U.S. government involved in aerospace. Mr. Sabathier has served in many roles in the aerospace industry, with the French Ministry of Defense, CNES, and Arianespace, where he was actively involved in the development of Ariane 5. Arianespace selected him to negotiate production contracts for Ariane launchers and later appointed him program manager for follow-on projects they financed.

G. Ryan Faith is an adjunct fellow and former program manager for space initiatives at CSIS. He was previously a CSIS intern working on international security issues and later the program coordinator for the Abshire-Inamori Leadership Academy. He has worked on publications on regional military balances, the Iraq War, the proliferation of weapons of mass destruction, and international trade. Currently, he consults on issues related to combat vehicle procurement. Mr. Faith holds a B.Sc. in engineering mechanics from the University of Illinois at Urbana-Champaign, as well as a B.A. in chemistry and mathematics (with minors in business administration and economics) from Rosary College (now Dominican University).

Ashley Bander is the program manager for space initiatives at CSIS. Ms. Bander worked previously for United Space Alliance and, before graduate school, was an intelligence analyst focusing on a range of transnational issues. She holds a B.A. in political science and international studies from Virginia Tech and an M.A. in diplomacy and international relations from Seton Hall University.